# HOW LONG HAVE WE GOT?

**BEATTY MEMORIAL LECTURES**

*Published by McGill–Queen's University Press*

THE CHANGING STRUCTURE OF
THE WESTERN ECONOMY
DOUGLAS COPLAND
1963

MONTREAL AND THE FUR TRADE
E. E. RICH
1966

THE BALANCE OF POWER
MAX BELOFF
1967

ASIA TODAY
TWO OUTLOOKS
HAN SUYIN
1969

RITCHIE CALDER

# HOW LONG HAVE WE GOT?

*Beatty Memorial Lectures*

MCGILL—QUEEN'S UNIVERSITY PRESS

MONTREAL    LONDON

1972

ISBN 0-7735-0150-9

Library of Congress Catalog Card No. 72-80883
Legal Deposit 2nd Quarter 1972

PRINTED IN CANADA
BY JOHN DEYELL LIMITED

# FOREWORD

---

Over the years, the Beatty lectures have brought to McGill and to Montreal the wisdom and scholarship of S. Radhakrishnan, Lady Barbara Ward, Sir Julian Huxley, Morris Bishop, Arnold Toynbee, Sir Douglas Copland, A. L. Rowse, E. E. Rich, Max Beloff, and Han Suyin. There could have been no more appropriate successor to this distinguished company than Lord Ritchie-Calder.

His subject was science and the way men use and misuse it. As you read his lectures, you will be struck by the way his grasp of the implications of science is matched by an experienced journalist's sensitivity to the public's concerns. You will be aware of an ability to combine sophistication of understanding with simplicity of expression that is rare and much needed today. Moreover, you will be conscious of the lecturer's deep humility, which came across so clearly to his listeners.

Lord Ritchie-Calder's assessment of the impact of science and technology neither disguises nor exaggerates the dangers we face; it neither underestimates nor inflates science's potential for good. He is a provocative and unfailingly sensible guide through such tangled and disturbing problems as how to preserve our environment, how to maintain our privacy, and, most fundamentally of all, how to survive.

McGill University is pleased and honoured to present his thoughts, through McGill–Queen's University Press, to a wider audience.

MICHAEL K. OLIVER
*Chairman*
*Sir Edward Beatty Memorial Lectures*
*Committee*

# PREFACE

THE Beatty Memorial Lectures which I was invited to give were billed as *Science and Social Change*. In the process of preparing them, and even more in the process of delivering them to responsive audiences, I began to feel that the title did not convey the sense of urgency which was building up.

Any analysis of the human predicament must remind us that science and technology form the economic and social dynamic of our time and that it is the application of science (human curiosity) through technology (human skills) which is piling up the problems and, by default of institutional competence, aggravating our difficulties. It seemed therefore that as a title *How Long Have We Got?* better reflected the manifest concern and the logic of the series.

I have dealt more fully with some of the ideas expressed in these lectures in *Man and the Cosmos, Evolution of the Machine, Leonardo and the Age of the Eye,* and *Men Against the Frozen North.*

I am abundantly grateful to my hosts at McGill University not only for the opportunity which they gave me but also for the tolerance and encouragement which they displayed. RITCHIE CALDER

*The Lord Ritchie-Calder of Balmashannar*
*The House of Lords, London*

# CONTENTS

---

# I

## SCIENCE AND INTERNATIONAL RELATIONS

---

IN SEPTEMBER 1971 it was announced that the Hot Line between the White House and the Kremlin (with Ottawa and London on the party line) was to be routed by communications satellite. The change, we were told, would improve the precautions against a nuclear war by accident. This was supposed to be re-assuring in a world in which the stockpiles of nuclear warheads already comprised the equivalent of 100 tons of T.N.T. for every man, woman, and child on earth, for each of the 3,600 million.

The cold comfort of the Hot Line was its tacit acknowledgement by the Super Powers, the United States and the Soviet Union, that whatever their postures on the Deterrent, they had ruled out nuclear war as an act of policy. That was the compensating good that came out of that nerve-wracking Rehearsal for Doomsday, the Cuban missile crisis in October 1962, when the Soviet rocket bases were detected and John F. Kennedy reacted with his declaration of 'quarantine', the blockade of Cuba to prevent the arrival of warheads. The Super Powers were horn-locked on the

brink of the Third World War, which inevitably
would have been a nuclear war. Fortunately the sun
still rises in the east and radio waves travel at 300,000
kilometres per second. Moscow is eight and a half
hours ahead of Washington, and at the eleventh hour
and fifty-ninth minute, Kennedy snatched from the
air a Khrushchev broadcast agreeing to dismantle the
bases. The U.S. task force was stopped on its way to
invade Cuba. That was the moment of truth for states-
men juggling with nuclear weapons as (in Clausewitz'
classical description of war) 'the extension of policy by
other means'. All the goobledegook of the nuclear stra-
tegists, 'first strike', 'second strike', and 'acceptable
casualties', became irrelevant. Decision-takers shrank
from the apocalyptic prospect of 'vetoing the evolution
of the human species', to use the phrase of that great
Canadian Dr. Brock Chisholm.

That, however, did not remove the risk of nuclear
war by accident. In the hair-trigger nature of the Deterrent, 'first strike' and 'second strike' are built into a
precarious system—a system which becomes more precarious every day. The essence of the system is Early
Warning which, in terms of intercontinental ballistic
missiles, gives fifteen minutes' notice, one way or the
other, of the impending arrival of multimegaton warheads. Once that warning (aggressive 'first strike') has
been received the only response is instantaneous re-
prisals (defensive 'second strike') which will incapaci-
tate the attacker. Under the now outmoded system

when Curtis LeMay's Strategic Air Command had manned bombers at the ready with warheads for delivery on Moscow, identification of 'first strike' depended on human monitors watching the blips on the radar screen. Two alarming incidents (of many lesser ones) showed the fallibility of the system. The first was when the blips showed a salvo of missiles heading for the heartlands of the United States and the airborne patrols were directed to Moscow to be recalled a few minutes short of Fail-Safe. That means that, beyond Fail-Safe, nothing, not even the voice of the President of the United States, could recall them, and the bombers simply became manned warheads pinpointed on the Kremlin. They were stopped because the 'salvo' was identified as a flock of Canada geese flying south. In the second case, even closer to the point-of-no-return and radio blackout, the blips were identified as radar signals, from the detection system, being reflected from the moon.

Those were the risks of human error but now, I am assured, such things cannot happen because the computer has taken over the unmanned missiles, although it beats me how the computer, dependent on pulses, can distinguish between Canada geese and a salvo of rockets. The last resort of human judgement is the President of the United States (or his opposite number in the Kremlin), who has fifteen minutes in which to press, or not to press, the button which, with its black box, he takes with him wherever he goes—to lunch with the Queen and the British Prime Minister at Chequers and, presumably, on his visit to China. He has a yes-no choice: to believe the computer or not.

Indeed, I understand that the refinements have gone further than that. The only value of the A.B.M. (antiballistic missile) is that it intercepts and destroys the intruding rockets. Thus it has to be automatically triggered off by the incoming identification signals. And, beyond that, still in terms of automation, those same signals will trigger off 'second strike', the instantaneous reprisal.

As part of the antiballistic game, the U.S.A. on 6 November 1971 carried out an underground test on the island of Amchitka in the Aleutian Islands. When the intention became known, there were public demonstrations in Canada, Peru, and Japan. Distinguished scientists of many countries, including one of the advisers to the U.S. President, made strong representations. This was to be a five-megaton bomb, that is, the equivalent in nuclear explosive power of five million tons of T.N.T. and 250 times as great as the bomb which destroyed Nagasaki, and it was to be detonated in an enclosed space, 6,000 feet underground, on the small island of Amchitka. This island, as was pointed out, is in the arc of the earthquake system which extends down the east coast of Kamchatka and Japan and down the west coast of Alaska, British Columbia, and California. This is a precarious area, about which far too little is scientifically known. Apart from the earthquake risks, a venting of radiation from such a test could have escaped into the polar system, which is a dynamo of our climate, and into the world's ecosystem, affecting the marine biology. This, therefore, was not the private concern of the United States; it was fraught with unpredictables. A century ago Claude Bernard, the famous French scientist, enjoined his colleagues: 'True science teaches us to doubt and in ignorance to refrain.' What that means is not that you do not progress, but caution in experiments demands that you should venture into the minefield of the unknown with a mine detector, making sure of every foothold. World scientists could not, on evidence available, feel assured that this was so in the Amchitka test.

The protests were ignored and the test proceeded. There was no serious incident (just a few seals) and the U.S. authorities got their evidence of the effectiveness of the antiballistic missile, but there could be long-term effects, in the earthquake system, which will be recognized only by hindsight as belonging to November 1971.

Another regrettable aspect of this affair was that it made a farce of the test ban. We had managed to restrain testing in the atmosphere (after the authorities, for years, had pooh-poohed the protests about fallout) but this partial ban was followed by underground testing of greater and greater magnitude. The Super Powers had transferred the dangers from the atmosphere to the lithosphere (while the French and

Chinese tried to catch up with their air shots). Moreover, the Amchitka test took place during the Strategic Arms Limitation Talks (SALT) in which the Super Powers were supposed to be seeking constraints. Coincidentally, U.S. reconnaissance satellites revealed some ninety holes in the ground in the Soviet Union which could be the silos for more missiles and, perhaps, new types of missiles. Thus we were back in the 'numbers game' in which each side blames, and outbids, the other in an auction of death. This escalation seemed unnecessary (to put it mildly) at a time when the Americans and the Russians both had around 2,000 launchers—land-based, submarine-borne, and bomberborne. The United States probably had twice as many warheads as the Russians ready for delivery. They had clustered them into Multiple Independently Targetable Reentry Vehicles (MIRVs for short) and tenheaded Poseidon submarine missiles. The Soviet also had hydra-headed missiles.

In such a situation the Hot Line becomes all the more urgent because it means that when there is a mistake—a faulty transistor in the computer, an error in programming, a wrong identification of a radar blip —the White House calls the Kremlin (or vice versa) on the Hot Line and says, 'Sorry about that multiheaded, multimegaton rocket which is going to vaporize you ten minutes from now. It was a mistake. Forget it politically.'

This verifiable cautionary tale of the Hot Line is the macabre reminder of how science has affected international relations. It embodies four epochal revolutions: the Atomic Age, the Communications Age, the Computer Age, and the Space Age, all compressed into the growing-up span of the university students of today.

Never in the whole history of mankind, since our ancestors first mastered fire, have so many stupendous innovations impinged upon society and never have the social instruments been so inadequate in dealing with change. The politicians, the diplomats, the financiers, the economists, and the social scientists concede the transcendental brilliance of the scientists and surrender as hostages to the technologists. Many scientists, the truth-seekers and the fact-finders, express their social unease in such things as the Pugwash Movement. Most technologists, functionalizing laboratory discoveries in systems and gadgetry, press on regardless. (Well, not quite regardless—not when tens of thousands of Ph.D.'s and graduate engineers find themselves unemployed in a society which their own efficiency has distorted.) Executives, the 'hunch-men' of classical business, find themselves drawing social security—fired by the computers they hired, since the most junior member of the board, armed with computer facts and figures, systems analysis, cost benefit, and the games theory, can discount his seniors' lifetime experience, innate shrewdness, and business second sight. Their quandary is that they cannot argue with the computer on which they spent all that stockholders' money.

According to Jacques Ellul, the French philosopher, technology is now a closed circle and the answers to the problems created by technology will have to be found within technology. On closer examination, it is not quite so absolute: technology is not a closed circle; it is a spiral which is spinning so fast that it looks like a closed circle; but it is still open-ended. Judgement and volition may still be able to function, but the options are being reduced, as in the parable of the Hot Line.

When the Charter of the United Nations was signed in San Francisco in June 1945, four people present should have known, but they did not understand, that they were legislating for a world that no longer existed. They were Truman, Stettinius, Attlee, and Eden. They were the only ones apprised of the fact that the atom bomb was about to be exploded. (It was, a month later on 16 July in the New Mexico desert where it made a crater which was to be the notch-mark in history from which the future of mankind would henceforth be dated.) They knew, although they did not realize its significance, because Truman had been let into the secret, but inadequately briefed, when Roosevelt had died two months before; Stettinius knew only by courtesy; Attlee knew it only as a bigger bomb; and Eden knew it only as the classified signals which crossed his Foreign Office desk. None of them knew how portentous it was nor had they considered fully its political as well as military consequences. (Whether the Canadians, who, of course, had been involved through the Montreal Project, had more insight or foresight, I do not know. I do know that under the notorious 'terms of co-operation' imposed by Dr. Conant from the U.S. side 'all matters concerned with the bomb and use of material' was to be withheld from the Montreal team.)

The Founding Fathers of the U.N. did not realize that the innermost secret of matter was to be released as an apocalyptic event. They did not know that the safebreakers were cracking the lock of the nucleus before the locksmiths knew how it worked. (Twenty-six years and billions of dollars later, the fundamental physicists are still trying to determine the structure of the nucleus.) None of them knew the possible biological effects. Years later Clement Attlee, who as Churchill's successor as Prime Minister accepted the atom commitment and concurred with Truman's decision to drop the bomb on Hiroshima, confessed: 'We knew nothing whatever at that time about the genetic effects of an atomic explosion. I knew nothing at all about fall-out and all the rest of what emerged after Hiroshima. As far as I know, President Truman and Winston Churchill knew nothing of these things either, nor did Sir John Anderson, who co-ordinated research on our side. Whether the scientists directly concerned knew, or guessed, I do not know. But if they did, then, so far as I am aware, they said nothing of it to those who had to make the decision.'* Incredible but true! In the single-mindedness of war, this was the physicists' bomb which they produced behind the sky-high fences of military security that debarred the biologists from knowing anything about it. Yet, in 1927, Hermann Muller had already demonstrated the effects of radiation on the genes, the factors of heredity, for which he was later awarded the Nobel Prize. Radiation mutation was textbook stuff but none associated it with 'just a bigger bomb', in Attlee's phrase.

Nor did the Founding Fathers of the United Nations, who were enlightened enough to create the Economic and Social Council, foresee the economic and social possibilities of the peaceful uses of atomic energy, and it was another ten years before the United Nations got around hopefully to considering the benefits which it might offer developing countries. Then in 1955 in the euphoria of the first Atoms-for-Peace Conference at Geneva, when we were talking about foot-loose energy from the nucleus being made available to economically underprivileged countries, one would have imagined that all a power-starved country had to do was order a packaged reactor and have it delivered. That promise has not been fulfilled nor has

* Francis Williams, *A Prime Minister Remembers: The War and Post-War Memoirs of the Rt Hon. Earl Attlee* (London: Heinemann, 1961), p. 74.

the premature prospect offered at that conference of thermonuclear energy harnessing, to industry, the energy of the H-bomb. That, apart from 'demobilizing the H-bomb and putting it into civilian dungarees', was a philosophically encouraging prospect because thermonuclear or fusion energy comes from the building up of atoms while fission reactors depend on the breaking down of atoms with all the waste problems of radioactive debris.

No one in 1945 saw the significance of another new dimension in human affairs—the crash program. It is also no exaggeration to say that, provided we avoid a nuclear war, the Manhattan Project which produced the nuclear bomb was as historically important as the bomb which it produced.

Otto Hahn, who had been a research student with Ernest Rutherford at McGill in the Dawn Age of the Atom, made the momentous discovery, in 1938, of uranium fission. This led to a chain reaction and the possibilities of the atomic bomb. In the year which followed his announcement over one hundred major, innovating, scientific papers appeared. Journals like *Nature* and *Science* were the bazaar and mart of the traffic in fundamental ideas triggered off by Hahn's discovery until, in 1940, western scientists applied a self-denying ordinance by which they withheld papers which might help the Nazis to produce the bomb. By December 1942 the chain reaction was being controlled, in Fermi's atomic pile at the University of

Chicago, to manufacture plutonium. Plutonium, like Uranium 235, is a fissile material, which, in uncontrolled instantaneous chain reaction, produces the nuclear explosion. The Manhattan Project, under military control in the United States, brought together hundreds of the greatest scientists, not only of the United States, but of the free world. It took their knowledge. It mobilized America's vast, intact technological resources. It secured a monopoly of Canada's supplies of natural uranium. It built a great plant at Oak Ridge, Tennessee, to separate fissile Uranium 235 from the natural uranium by diffusion. This process involved an incredible amount of wire for the electrical equipment, and when copper ran short the Project commandeered $200,000,000 of silver from the Federal Bank reserve and converted it into filament. The Project built a huge production plant at Hanford in the State of Washington to manufacture the alternative fissile material, Fermi's plutonium. J. Robert Oppenheimer was put in charge of the bomb development centre at Los Alamos, New Mexico, where he assembled his scientific colleagues and the bomb materials. On Monday, 26 July 1945, at 5.30 A.M. to the precise countdown second, the laboratory discovery of 1938 exploded with cataclysmic violence. Man had released the enormous energy of the nucleus. Thus, by means of a crash program, with no limit on money, and with the mobilization of brains and skills, the time-calibration of change had been reduced from centuries to decades, from decades to years, and from years to months.

A similar thing happened with electronics, the basis of the Communications and Computer Revolutions. Practical electronics in the form of radio and television, of course, were well advanced at the outbreak of the Second World War. The crash program on radar, however, accelerated the whole process. Sir Robert Watson-Watt's surface-based detection system, which defended Britain against the sorties of the *Luftwaffe*, had become airborne. This result was possible because of the invention of the cavity magnetron by Randall, Boot, and Sayers at the University of Birmingham. This hunk of copper, with cavities like the reeds of the pipes of Pan, enabled electrons to go round in increasing crescendo until a pulse was released equal to the power of a broadcasting transmitter. Sir Henry Tizard was able to take the cavity magnetron in his briefcase to the United States. In the same way, other components were miniaturized. The thermionic valve made of very tough glass was reduced to the size of a vitamin capsule. By these means bombers were able to send radio signals from aircraft to the ground and have them reflected back on to a cathode ray screen in the cockpit, so the details of the bomber's target were illuminated by invisible rays. This was $H_2S$ or 'Blind Bombing'. From such developments in miniaturization, electronics entered a new phase. If one could imagine a computer made out of prewar thermionic valves, it would have had to be housed in a building as big as an Expo pavilion. But with the subsequent development of transistors, requiring low power and no cooling system, it was possible to devise the kind of things which we have in hundreds of varieties today. Here we have another example of a crash program telescoping the time required for development.

The computer and automation of which it is the mastermind were both the result of such time-telescoping. Machine computing goes back to the abacus of the ancient Chinese—the bead frame—and graduates through Blaise Pascal's cogged-wheel adding and subtracting device in 1642; through Charles Babbage's calculating engine (with a punch-hole memory) of 1844; and D. D. Parmalee's primitive cash register of 1850. No one today, however, imagines that a computer is just an electronic abacus. True, it can do prodigious calculations at fantastic speed; it can do in seconds sums that would take a prize mathematician a lifetime with pencil and paper. But it can also simulate the logical faculties of the human brain and can be vested with a memory of superhuman capacity. Using an eight-coloured laser beam, 100 million bits of information can be stored on a square of photographic film no bigger than a postage stamp. An entire library of 20,000 volumes can be stored on a piece of nickel no bigger than a book page. With cryogenics exploiting the behaviour of certain metals at, or near, Absolute Zero (minus 237 degrees Celsius), it would be possible to store all the information of all the libraries in the world in a casket no bigger than the human cranium. Such miniaturized information can be retrieved by the computer from its memory and printed out as legible extracts. Moreover, the computer is combined into the systems which we call automation and cybernetics. What this means is that machines have today not only replaced muscle effort but have acquired the equivalent of the human senses, the central nervous system, and some of the faculties of the human brain. The microphone is more sensitive and more tireless than the human ear; the photoelectric cell is more sensitive and more tireless than the human eye; and far more sensitive than the human touch are machines which can operate to precisions of a millionth of an inch. Coupled with these is 'feedback', which is like the nervous system. It automatically responds, makes adjustments, and corrects mistakes without human intervention. Those senses, as we have seen in the Space Program, can reach out their sensitive nerve-endings to the planets and can telemeter their responses back to earth.

Like the release of atomic energy, the Space Program derived from military violence, as the British, at the receiving end of the German V-2 rockets, knew to their cost. In the immediate postwar years, the Soviet Union and the United States acquired leading German scientists (for example, Wernher von Braun) who were responsible for the V-2s and embarked on rocket missile programs.

Then in 1957 we had the International Geophysical

Year, one of the greatest scientific enterprises which
the world has ever known. The scientists of over one
hundred nations cooperated in a joint effort to study
Planet Earth, its envelope of atmosphere and the cosmic forces which impinge upon it. One of the agreed
I.G.Y. projects, in addition to the one-time-go rocket
probes into space, was to put artificial satellites into
orbit around the earth to look inwards and survey the
planet and outwards to intercept cosmic rays, and so
on. It began as a modest intention. The Americans
and the Russians agreed to supply and launch the
observational satellites, with instruments but unmanned. It was naturally assumed (even, I imagine, by
the Russians) that the Americans, with their great
technological resources, would be the first to launch
their satellite, which, incidentally, was to be not much
bigger than a football. But on 4 October 1957 the
Soviet satellite, Sputnik, was launched. It weighed
eighty-one kilos and carried instruments to determine
and relay back to earth data on temperatures and pressures.

The bleeping of that first Sputnik caused a crisis in
international relations. The size of Sputnik and its
orbit raised an outcry in the United States about 'the
missile gap'. The U.S. armed services insisted that the
powerful launcher showed that the Russians had intercontinental ballistic missiles since they could spare one
for Sputnik, which meant that they could reach from
Russia to the Western Hemisphere. We now know that

there was no missile gap—no adverse numerical disparity between Soviet and American I.C.B.M. capacity. But the outcry led to the Space Race, culminating in another crash program—Man on the Moon by 1970. This goal the United States achieved, ahead of schedule and at a cost of $40,000 million, on 21 July 1969. It was another example of how, if you pay the money and mobilize the manpower, science and technology will give you what you ask for.

The Space Program also contributed communications satellites and surveillance satellites. The first made it possible to relay television programs so that we could see, instantaneously, events happening on the other side of the world. And they provided radiotelephone channels, including the Hot Line. The surveillance satellites, in orbit above the atmosphere, provided pictures for the weather men and also means of studying vegetation, even detailed crops, and, by extremely delicate instrumentation, geological deposits still unprospected on the ground itself. They could police' pollution in the atmosphere and the sea. But they also had a profound military significance. We were made aware of the fact that high-flying aircraft could get detailed information about military installations within a country (the shooting down of the U-2 spy plane over the Soviet Union had broken up the Paris Summit Conference), but satellites in orbit do that job more thoroughly and continuously.

That brings us back to the Deterrent, alias the

Balance of Terror. With increased efficiency of aerial and satellite surveillance systems and with the target accuracy of nuclear weaponry, permanent land bases became vulnerable. Just as U-2 photographs showed the building of roads and the pouring of cement for launching pads in Cuba, the activities associated with missile installations in the heart of the United States or the Soviet Union are revealed and become a fixed target. The answer is mobility or concealment. The opaque depths of the sea provide both. Polaris submarines and other submersibles keep the launchers on the move and hidden from the camera-eye. When the Russians and the Americans agreed on a pact not to emplace or implant weapons of mass destruction on or in the ocean bed, and invited other nations to ratify, it was no great concession—just a nice gesture. Obviously, there is little advantage to anyone to have fortresses on the ocean bed if the water medium can keep the potential targets on the move.

Thus activities in Outer Space (the satellites) led to intensified interest in Inner Space (the oceans). Research and technology for operating at great depths were speeded up. When defence departments, through their appropriations, take care of research and development costs, the time-scale of technological innovation becomes radically different. The spin-off from an expense-no-object military program can be the know-how of civilian operations. The missile program became the satellite program; the bleeps of Sputnik became the telecast of Man on the Moon in 1969, twelve years later.

With the cutback in the Outer Space program in 1970, the military-industrial complex of the United States was looking for diversification and for the employment of their know-how and manufacturing capacity for purposes of Inner Space.

In terms of international relations the oceans have become a major preoccupation. Historically, the oceans, all 140,000,000 square miles of them, were the waters which separated the continents and islands and provided the thoroughfares for trading ships and, in war, the battlegrounds for navies. Periodically, nations with a sense of naval supremacy, for example, the Portuguese, the Spaniards, and the English, would claim dominion over areas of the open, or high, seas. The English jurist, John Selden, in 1635, expounded the legal doctrine of *Mare clausum* in which he asserted that 'the sea by the law of nature or nations is not common to all men but capable of private dominion or property as well as the land'. This, however, did not prevail against *Mare liberum* in which Hugo Grotius, the Dutch jurist, in 1609, had embodied the doctrine of the Freedom of the Seas, qualified only by the practical need of a coastal state to exercise some jurisdiction in the waters adjacent to its shore, which, in the eighteenth century, was defined by van Bynkershoek as the distance which could be protected by land-based cannon. This range (overambitiously for the weaponry of the time) was defined as three miles.

Beneath the surface of the seas on which ships had the right of free passage, there were fish. As fishing vessels extended their ocean-going capacity and could reach fishing grounds farther and farther from their own shores, coastal nations sought to safeguard the livelihoods of their own fishermen within territorial waters and to extend the limits of those waters. These attempts took the form of armed protection against foreign competitors as well as a marine constabulary to ensure good fishing practices by their own nationals. Furthermore, the navies of maritime nations, in the common interest of all, provided a form of collective security against piracy on the high seas.

The Law of the Sea was thus, for centuries, concerned with a liquid medium and no more. All that was changed by the Truman Proclamation of September 1945, which asserted jurisdiction upon the submerged land stretching from the shores of the United States to the outer limits of the Continental Shelf. This, in origin, was a domestic matter: the existence of offshore oil was discovered in the Gulf of Mexico and the technology for gaining access to it had been evolved. The President was, in fact, asserting the claims of the Federal Government against the claims of the adjacent states of his own republic. It was, however, a proclamation of momentous importance. It set the example for the Argentine, which in 1946 declared its sovereignty upon a 200-mile strip of sea bed. The Law of the Sea Conference of 1958 recognized the principle that submerged territories belong to the coastal nations adjoining them and come under their jurisdiction. The outer limit was ambiguously defined, being the 200-metre isobath (sea depth) or beyond that as far as the depth of water allows the exploitation of sea-bed resources. Since then we have had various attempts to define the submerged provinces—the edge of the Continental Shelf, the limits of the Continental Slope, which falls away from the shelf, and even the Continental Rise, the spill-off of continental materials that stretch to the abyssal plains. This 'creeping jurisdiction' is something which has to be contained. By the most restricted definition, that of the Continental Shelf, the submerged territory has added to coastal nations 22.5 million square kilometres, which is four and a half times the surface of Europe outside the Soviet Union and more than the area of Africa, south of the Sahara.

Beyond the limits of national jurisdiction, however that may be ultimately defined, there is still a vast area of the drowned land of the earth which the nations of the United Nations have agreed is the 'common heritage of all mankind'. This agreement raises fascinating new issues. For one thing, we are called upon to define for the first time what is meant when the preamble of the United Nations' charter said 'We, the peoples . . .'. It did not say 'We, the governments'. Nor 'We, the

High Contracting Parties'. It said 'We, the peoples' who now number over 3,600 million individuals and who, regardless of race, creed, or geography, own a valuable property—the real estate of the sea bottom.

Once we thought of the ocean bed as a desert with no wealth except sunken galleons and bullion ships. It was a place where we could safely dump all our trash, including dangerous poisons and radioactive waste. Now we know it is nothing of the sort. Apart from Sputnik and starting the Space Race, the International Geophysical Year confirmed the extent and potential economic value of manganese nodules on the ocean bed.

The ocean has been described as a liquid mine. It contains in solution, in suspension, or in deposition all the earth's elements. The minerals are disposed in the ocean by various mechanisms. They come from the mantle through volcanic fissures and vents. They come from meteorites for which the oceans offer a much bigger target than the land area, receiving something like four million tons of cosmic debris per year. Mainly, however, they come from the crustal rocks of the land surface. The weathering of rocks, the scouring by rains, the action of winds, and the 'open-cast mining' by streams and rivers carving their courses from their watersheds to the seas make their contribution. The waves themselves are a form of hydraulic mining, undercutting cliffs, with their mineral formations, and surf-grinding the hardest rocks. On conservative estimates the oceanic waters contain fifteen billion tons of copper, seven thousand billion tons of boron, fifteen billion tons of manganese, twenty billion tons of uranium, five hundred million tons of silver, and ten million tons of gold; and all other elements in proportions of millions and billions of tons. Diamonds, platinum, placer gold, and tin are dredged up from the sea floor in relatively shallow areas. Ocean rigs drill into the submerged Continental Shelf to get natural gas oil and molten sulphur. (Today a sixth of the world's oil comes from offshore drilling.) Bromine, common salt, and magnesium are separated from the brine. But the others present insuperable problems of separation from the liquid. (The Germans tried to electrolyze gold from the oceans to pay World War I reparations and failed.)

The sea itself, however, by its own alchemy and by the leisurely processes of eons of time, has made substantial conversions. They take the form of 'manganese nodules', which is a misleading term because with the manganese is incorporated other valuable minerals. These nodules were first brought to the surface a century ago by the British oceanographic vessel *Challenger*, which dredged them from the deep parts of the Atlantic, Pacific, and Indian Oceans. From the turn of the century, in spite of the fact that the *Albatross* expedition found that the nodules covered an area of the eastern Pacific larger than the United States, the nodules remained laboratory curiosities. Then in

1957/58, the I.G.Y. surveys around the world showed that the nodules were, in composition and extent, a major source of economic minerals. 'Economic', of course, means that they are profitably recoverable. As has been pointed out, the technology was advanced by the research and development of the deep-water military strategy. Depths and pressures are no longer regarded as obstacles. Materials-technology is already far enough advanced to promise manned vehicles (even habitats) at great depths, although the Man in Depth will be merely incidental to most of the operations involved in extracting resources from the ocean floor or its subsoil. One example of contradicting the seemingly impossible was provided by a recent achievement of *Glomar Challenger,* the survey ship, which drilled into the bottom through 16,000 feet of water and withdrew and replaced the drill. It was like rethreading a needle with 16,000 feet of thread. Indications now are that there is likelihood of oil at depths well beyond the Continental Shelf, so there are possibilities of deep-sea oil drilling in the property of 'We, the peoples'. Industrialists who have their eyes on the nodules reckon that they can be recovered by scraping and trawling the ocean floor, a process equivalent to open-cast mining. Thus the industrialization of the deep oceans is imminent.

This state of affairs makes more urgent the need for an Ocean Regime to regulate and license such activities, to prevent the expropriation of the property of

'We, the peoples', and to prevent pollution through reckless methods. The matter is in continuing debate in the United Nations Standing Committee on the Peaceful Uses of the Ocean Bed, but it is difficult to convince people of the genuine urgency or that un-regulated recovery of ocean minerals will presently be competing with those of the land to the ultimate economic disadvantage of developing countries whose assets are such raw materials.

The present position is that 'common heritage' is, in lawyers' language, either *res nullius* or *res communis,* the first meaning belonging to nobody and the second belonging to everybody. In other words, it could be a free-for-all.

Assuming that the nationals of a country were to go out and exploit the mineral resources to which, under the common heritage principle, all governments have renounced claims, and to which there is no authority to grant them concessions, and assuming that those nationals were to establish (as is possible) a man-made island, an ocean platform, above the site, would the state of which they were nationals provide them flag protection to sustain the claims against all-comers? What would happen if the 'comers', trying to muscle-in, were similarly to claim the flag support of their state? This would be tantamount to military expropriation of our ocean property and would more than likely lead to armed conflicts. But consider the role of multinational corporations that might combine, with no nationality of origin, in an enterprise of this size and novelty. One of the present characteristics of multinational corporations is that a foreign concern with technological know-how and managerial expertise combines with the nationals of a country whose resources or markets they are seeking to develop and, by this incorporation, secures the protection of their enterprise. But in the deep sea there is no means of doing this. One conceives of a consortium of firms combining their capabilities and their technology in multinational anonymity with neither a flag of state nor a flag of convenience. Through what constabulary could they sustain their activities? The answer is 'Their own'. They could provide their own security force, as firms now do to protect their plant against burglary, sabotage, or industrial espionage. It is possible to foresee private navies maintaining or extending the claims of corporations like maritime feudal barons.

A legal Ocean Regime is essential and it cannot be just a cartel of governments, each of which would be an interested party in the obtaining of licences. So we have now an opportunity, which I welcome, for creating a new kind of institution, a People's Trusteeship, to regulate concessions and see that the proceeds, whether in royalty or rent or shared dividends, are distributed for the benefit of the developing countries, land-locked as well as sea-bound, to redress the disparities of today.

The industrialization of the ocean bed, with its risks of pollution, reminds us of another 'common heritage', the biosphere on which all life on this planet depends—human, animal, and plant life. Belatedly we have come to recognize that impairment of the environment is no longer a local or national matter or a legitimate price to pay for private profit or national prosperity. In industrial Britain in the booming nineteenth century it was said 'Where there is muck there's brass'—where there is soot and slag heaps there is prosperity. What we now properly call the quality of life was ignored and so were the social costs. Those were disregarded in the interest of competitive prices. Britain was disfigured by the wealth it was supposed to be creating.

Now, in a world in which multiplying people are making multiplying demands for goods which create or become waste products, airborne and waterborne pollution is an international problem of ominous proportions. The object of the Stockholm Conference of 1972 on the Environment was to determine the state of affairs and the measures to be taken, but we are discovering how relatively little we know scientifically. The fashionable word now is 'ecology', which the dictionary describes as 'the mutual relations between organisms and their environment', but how much scientific content have we yet given to it? We call it the Great Chain of Being but how much do we know about the links in that chain? We are concerned about the manifest destruction of our living waters and the drastic threats to the oceans themselves but graduates in marine biology in Britain cannot get jobs. We talk grandiosely about 'the environment' but shrink fastidiously from the subject of sanitary engineering. We mouth words like 'eutrophication' but do not associate it with our own bowel movements. We clutch at words like 'recycling' but do not recognize that, most of the time, it is paying people to take out what they never should have put in. In Britain we can boast that we have reduced the sulphur at ground level to safe proportions, while the Scandinavians complain that their forests are being destroyed by the sulphur we have got rid of from our tall chimney stacks. Thus, internationally, we throw our wastes into our neighbour's backyard. If we are going to make our concern about the environment as urgent as it demands, we should be mobilizing the world's scientists in a crash program more meaningful than putting man on the moon.

Pierre Auger, the famous French scientist, reminded us that 90 per cent of all scientists and research workers that ever lived are still alive. Which is a way of saying that most of the great discoveries in the history of mankind have been made within the past fifty and mainly the past thirty years. Add to those discoveries the fantastic achievements of technology which took the laboratory discoveries and converted them, and one sees the extent of the Scientific and Technological Revolution which, since World War II, has given us the

Atomic Age, the Computer Age, the Communications Age, and the Space Age, each as epochal as the Bronze Age, the Iron Age, the Renaissance, and the Industrial Revolution. In the world of international relations another upheaval has been happening at the same time—the Revolution of Rising Expectations. This is a component of the Communications Age.

The winds of change to which Mr. Harold Macmillan referred when he was British Prime Minister in a disappearing empire were, in fact, etheric. They were radio waves. The aspirations of independence might be innate but what produced the epidemic of independence was the universality of radio. Independence movements found a common identity in their struggles against colonialism. The awareness of the success of one movement encouraged others. The time-scale of change diminished rapidly. When, as adviser to Lord Boyd Orr, the first Director General of F.A.O., I took the oath as an international civil servant (albeit temporary) in 1946, I swore that oath in front of the flags of 51 nations. Today the membership of the United Nations is 128, the majority being new states that have emerged from colonial status. Think of the long haul to self-government in Asia and the quick reactions to independence in Africa. It was inescapable. The events within a country reverberated in the councils of the United Nations and the discussions within the United Nations reverberated back into other countries. When I was young, Edgar Wallace, in his books about Sanders of the River, had a character, a stiff-upper-lipped British District Commissioner in West Africa. When the Ogawoga of Bogowoga got out of hand Sanders would get into his paddle steamer and puff-puff up the river, clip the Ogawoga over the ear with his swagger stick and puff-puff down the river again. That would have been the end of the incident. He would not have bothered to put it in a dispatch to the Colonial Office. Nowadays, there would be helicopters, trans-Atlantic calls to the U.N., and complaints in the General Assembly.

The aspirations to political independence are not the only thing fostered by radio communications. So are expectations of material betterment. The advanced countries boast of their great achievements—men on the moon, roads crowded with the latest cars, cures for this and that, abundant food in three types of dressing. And all they do is to remind people in developing countries that they are still sick and hungry and miserably poor.

The awareness of change is everywhere—and not just political change. I have been on the *alto plano* of South America where the dispossessed heirs of the Incas, beneath the gleaming bowls of the rocket trackers, were listening to astronauts gossiping on the threshold of Space. I have been in the Himalayas when the sherpas of Everest were listening to accounts of *Nautilus* going under the polar ice. I have seen Congolese in the rain forest of Central Africa listening to the news bulletins and tapping out the message on talking drums. I have been in the longhouses of Borneo with young Dyaks listening to pop music by courtesy of the Voice of America. I have been out with Eskimos in the Canadian Arctic when they were carrying geiger counters on their dog sleds to look for uranium as well as portable radio-receivers to tune into the metal-market prices to see whether it was worth while looking for uranium.

The Revolution of Rising Expectations, as well as the frustrations of those expectations, is a message that has got through to the underprivileged, and then we wonder why they are in a state of unrest and why governments that fail to provide promised satisfactions are unstable.

The dilemma of our times is summed up in the parable of the Hot Line. Science has encompassed knowledge, and technology has accomplished feats which have given man powers to determine his own destiny. By speed of transport and communications he has shrunk his habitat to the dimensions of a small planet round which an artificial satellite can circumnavigate sixteen times a day. He has reduced it to a neighbourhood in which no place is more than a few hours distant by jet propulsion, a few minutes away by intercontinental ballistic missile, or split seconds away by radio. He can stand on the moon and recognize the proportions of a world on which over 3,600 million people today and twice as many by A.D. 2000 have to contrive to live together. He has wrested its secrets from the atom. He has simulated some of the faculties of his own brain. And he has embodied all this in a system, the Hot Line, the confession of human frailty.

## II

## SCIENCE AND HUMAN RIGHTS

---

SCIENCE ITSELF is a human right—the fundamental right to know. Natural curiosity is the attribute of *Homo sapiens*, Thinking Man, in whom perception, which he shares with the animals, is combined with imagination.

Science, therefore, is the everlasting interrogation of nature by man. The interrogation began when Ancestral Man overcame his superstitious wondering and began to ask 'Why?' (And, in the childhood of mankind, there was no teacher to snub that curiosity by saying, 'You cannot understand that until you learn all about kinetics and dynamics'.) Early man noticed an object or an event and made a mental note; thus he became an observer. He looked again to make sure; thus he became a fact-finder. He speculated why, for instance, the sun and planets seemed to move in a repetitive way against the background of the firmament and in relation to the pattern of the stars; thus he became a theorist. He searched and re-searched, verifying his observations; thus he became an investigator. He began to assemble his ideas and to relate what he saw to other forms of experience; thus he became a natural philosopher. He exchanged his observations and ideas with others, who might accept his facts (from their own observations) but dispute his interpretations of the facts; thus there developed the dialectic—the art of reasoning about matters of opinion and the discrimination of truth from error. That was the beginning of the scientific debate that distinguishes the accretion of experimental findings from the hypotheses and theories based on those findings.

Science *per se* ascribes no moral values. The philosopher David Hume argued for a radical distinction between the 'is' and the 'ought', between fact and value. Science is the 'is'. Moral philosophy must provide the 'ought'.

The other self of *Homo sapiens,* Man, the Thinker, is *Homo faber,* Man, the Maker. Like man, his fellow creatures had a set of inherited instincts, their built-in experience, like the dam-building beaver or the nest-weaving birds. The more intelligent might spontaneously improvise (as a chimpanzee will contrive an implement to reach a banana beyond his reach), but apart from the instinctive preservation of their species these creatures have no grasp of the future. Man evolved differently. He had foresight. He could plan ahead and to his great benefit developed language, the means of passing acquired knowledge from one generation to another. Thomas Carlyle, inadequately, called man a 'tool-using animal'. Benjamin Franklin, more properly, described man as a 'tool-making animal'. The distinction is important, and if Franklin had added 'with foresight', his definition would have been complete.

The combination of perception, memory, imagination, logical foresight, and finger skills enabled *Homo sapiens* to survive as a species, in competition with his predators and in a hostile environment. The kind of knowledge which we now call 'science' activated by what we now call 'technology' asserted another human right—the right to survive.

Leonardo da Vinci, four hundred years ago, declared 'Science is the captain; practice the soldiers', by which he meant that no technician and no artist should be continually improvising and learning by mistakes when he could be captained by scientific laws revealed by measurements and proved by previous experiments. He made the relationship of theory to practice clear in one of his injunctions to himself: 'When you put together the science of the movements of water remember to put beneath each proposition its application so that such science may not be without its uses.' He rebutted contemplative science (which Plato had glorified in the statement: 'Every soul possesses an organ, the intellect, better worth saving than a thousand eyes because it is our only means of seeing the truth') and St. Augustine's injunction: 'Go not out of doors. Return into yourself; in the inner man dwells truth.'

What we call the Renaissance, to which Leonardo da Vinci supremely belonged, was the Age of the Eye, the pushing back of the cowl of cloistered intellect and the venturing forth to look at things. For centuries, the eye, because it was the subject of optical illusions, was the most distrusted of the senses. Seeing was not believing. Scholars were enjoined that the only sense to be relied on was the sense of touch. To accept anything one saw, one must confirm it by touch. Images could be seen in positions where touchable objects did not exist. A reflection was not tangible, like the three dimensional person it reflected, and, therefore, was a deceit of the eye. The dominating and dogmatic distrust of the sense of sight and, conversely, the reliance on cerebral insight delayed for centuries scientific achievements that only required external recognition of self-evident facts. The right to find out had been fettered by the obscurantists.

Then, four hundred years ago, as far as western culture is concerned, the eye opened, like the aperture of a camera lens. Sight became the paramount sense. Men saw things and saw their relevance. Some saw the raptures of nature in radiant colours. Some saw the elementary facts of nature with prosaic precision. Some saw the jig-saw pictures and how they could fit. Perceptual Man became Conceptual Man, putting apparently disparate things and ideas together. Inquisitive awareness became imaginative construction. In the common vision of the artist and the scientist (embodied in Leonardo) perception and conception were shared by the recorder and the inventor, the dreamer and the maker, the thinker and the doer.

Sir Francis Bacon reaffirmed the scientific method by insisting on experimental investigation. He himself could scarcely be regarded as a laboratory scientist (although he died as a result of an experiment in which, anticipating refrigeration, he went out to stuff a goose with snow and died of pneumonia). His writings (*Advancement of Learning, Novum Organum,* and *New Atlantis*) spell out what are now the widely accepted principles of modern science, rejecting the *deductive* or thinking-off-the-top-of-the-head principle, in favour of the *inductive* or take-off-your-coat principle. He insisted that the man of science must *observe* and choose his facts; he must form a *hypothesis* that links them together and provides a plausible explanation of them; and he must carry out numerous checks and repeated experiments to support or deny his hypothesis. Trained as a lawyer, Bacon applied to science the laws of evidence and the burden of proof.

The scientific method as applied today is an exacting discipline demanding that a man of science lay aside all his preconceptions. As Claude Bernard enjoined his colleagues, 'When you enter your laboratory, put off your imagination as you take off your coat but put it on again as you do your coat when you leave the laboratory. Before and between whiles let your imagination wrap you around; put it right away from you during the experiment lest it hinder your powers of observation.'

The man of science strives to attain the perfection of reducing everything to precise measurements. Sharing the beauties of the rainbow with the poet, he will nevertheless spell it out in angstrom units of wavelength. His olfactory nerves and tastebuds will help him enjoy the bouquet and flavour of wine but his stern mistress, Science, expects him to express it by chromatography in terms of aromatics. He may immerse himself in beautiful music but to make its sounds meaningful he will reduce it to pitch, frequency, and amplitude of sound waves. Exact science is quantitative and not qualitative.

The word 'scientist' did not appear in English, or, as far as I can determine, in any other language until 1840, and as late as 1890 the London *Daily News* was still deploring it as 'an American innovation'. H. G. Wells always preferred 'man of science' to 'scientist'. The distinction is important because the '-ist' meant the arrival of the specialist, and the development of specializations, each with its own self-propagating terminology. Previously, a man of science had meant a person of learning, especially inquisitive about nature but capable of sharing and explaining his findings with other educated men. Today terms have become more and more cryptic and exclusive.

Another process gained ground—what A. N. Whitehead called 'the greatest invention of the nineteenth century, the invention of the method of invention'. This was technology, the systematized transfer of knowledge from the laboratory bench to the factory floor. Thus, in the past century there has emerged a hierarchy of science: 'pure', or academic, science seeking knowledge for its own sake; 'oriented', or fundamental, research within a given frame of reference; 'applied', or programmed, science, which is research that is expected to produce foreseen results; and 'technology', which is the transfer into practice. Within such a hierarchy the practitioners might be distinguished as the Makers Possible, the Makers to Happen, and the Makers to Pay.

Today, the 'apartheid' is breaking down. The theoretician adjusts his parameters to experimental evidence, the experimentalist responds to the theory, and the technologist reacts to both in creating his 'hardware'; and the 'hardware' itself services the experimentalist and the theoretician. Through this feedback system, scientific progress has accelerated to a fantastic degree. The volume of scientific information has become a Niagara and natural philosophy has been swamped by experimental results. In terms of moral philosophy, the 'is' and the 'ought' rarely sit down together in the same room. We talk about 'the gulf between the Two Cultures', and lament the fact that science and the humanities are barely on speaking terms. It is true that, with the fragmentation of science into specialities and the proliferation of concepts and terms, there is difficulty in communication, but it is also true that there has been more effort on the side of science than of the humanities to make contact.

The fundamental right of finding out has to be respected and safeguarded. Unlike the old cartographers, we cannot write of unknown areas 'Here be dragons' and delimit Forbidden Territories. The medieval church tried it. But we know that there were candles in the dark, even in the Dark Ages. Discoveries were being made; the constraint was on their disclosure. The 'sublime Leonardo' was not the stuff that martyrs are made of: he was not going to incur the wrath of the Inquisition; nor would he invite the wrack or, like his exact contemporary, Savonarola, the burning faggots of the stake. Nevertheless, he pursued his researches and in his mirror, right-to-left handwriting he wrote *Il sole nó si move*, 'The sun does not move', a century and a half before Galileo. This heresy dethroned man from the centre of the universe and challenged his egocentricity as God's special creation. Leonardo, from his direct observations of the marine shells of inland Tuscany, disputed the story of the Flood, maintaining that they showed that the seas had once covered the area and not a rain deluge. He studied the nature of the rocks and rebutted the idea of a six-day Creation. He avoided the consequences by pretending to be the devil's advocate (and that was all right provided the devil did not seem to win the argument) and would write 'the adversary says' when he meant Plato, or Aristotle, or Galen, or some sanctified authority of the Church. When citing some forbidden experiment he would write as an injunction to an anonymous 'you', 'You will do so and so', when it was something he intended to do, or had already done, himself. He carried out the dissection of some thirty-seven human cadavers, against the prevailing Church ban. The determined investigator will defy any restriction. The sanction today, in these days of vastly expensive equipment, would be to withhold funds so that the impossible would take a little longer while ingenuity substituted for hardware. But curiosity would still prevail.

Anyway, suggestions of a moratorium on science are nonsensical. The kinds of knowledge we call 'scientific' are extremely diverse. They extend from the nuclear particles to the mental processes; from the birth and death of stars to the migration of birds; from the sea bottom to the expanding universe; from ultra-microscopic viruses to earthquake convulsions; and from the structure of crystals to the processes of aging. They deal with the macrocosmos and the microcosmos. They deal with the secrets of matter and the secrets of life. Such innumerable and endlessly diverse topics cannot be brought under any one formula.

What people really mean when they talk about a moratorium on science is 'putting the brakes on the *applications* of science'—not the finding out but the use, misuse, or abuse of what is found out. They are talking about constraints on technology. And this is where the 'is' and the 'ought' have to be reconciled.

The late Sir Richard Gregory, editor of *Nature*, one of the most socially responsible scientists I have ever known, used, between the two wars, to defend vigorously the 'innocence of science'. He would point out that on the same page of a chemical reference book one found juxtaposed chloroform and mustard gas. The one was regarded as beneficial to humanity and the other as a devilish weapon of war. The chemists first made chloroform in 1831, as a laboratory curiosity. Sixteen years later a medical practitioner, a clinical technologist, James Y. Simpson, after self-experimentation, used it as an anaesthetic. The discoverer of mustard gas was merely studying and isolating the interesting properties of that time-old condiment mustard. He found that fraction which made it useful as a poultice. This blistering factor was to be sublimated (save the mark!) into the vesicant 'Agent Q', which in minute quantities will burn and blind and, if inhaled as an aerosol, will kill. As Gregory used to say, it was the others who took the findings of science and turned them into destruction. And that was substantially true except that, in wartime, scientists themselves were patriotically prevailed upon to become the technologists of new weapons.

Nevertheless, until the atom bomb, Gregory could fairly acquit pure, or academic, scientists of premeditated evil. In the case of the bomb it was not the military, nor the politicians, nor the merchants of death who took existing scientific knowledge and perverted it. Scientists were responsible for converting the fundamental discovery of uranium fission by Otto Hahn into a cataclysmic weapon. The recognition of the chain reaction to which the discovery lent itself suggested possibilities of an explosion orders of magnitude greater than anything possible from chemical reactions.

Einstein had propounded in 1905 the equation $E = mc^2$, energy equals mass multiplied by the square of the speed of light, which means that matter can be converted into energy. It was to Einstein, the lifelong pacifist, that Edward Teller, Eugene Wigner, and Leo Szilard went in July 1939, within nine months of Hahn's discovery, and persuaded him to write to President Roosevelt, drawing his attention to the fact that 'this new phenomenon would lead to the construction of bombs . . . extremely powerful bombs'. Years later, appalled by the prospects which this proposal had produced, Einstein was to try to persuade President Roosevelt not to use the bomb, then in production. (Roosevelt never got the letter.) And later still, after the horrors of Hiroshima, he was to try to stop President Truman's H-bomb: 'If successful, radioactive poisoning of the atmosphere and hence annihilation of any life on earth has been brought within the range of technical possibilities. . . . Every step appears as the unavoidable consequence of the preceding one. In the

end there beckons, more and more clearly, general annihilation.' Many, indeed most, of the great scientists
who had been involved in the proposals and their ful-
filment suffered the same moral revulsion. A sense of
guilt was built into science which had not been there
before. It led to the great Pugwash Movement, which
one might call 'the ecumenical of science', bringing
together scientists from all over the world and, beginning with the nuclear physicists burdened with their
involvement in the Bomb, bringing together many
disciplines to consider their responsibilities to mankind. Pugwash, Nova Scotia, was the birthplace of
Cyrus Eaton, the layman, who first called the conference and brought together West and East scientists
whom the politics of the Cold War had divided.

Just as in that spectrum which extends from pure
science through technology to technics it is difficult to
draw lines of separation, so is it invidious to distribute
blame (if and when there is cause for blame) between
the Makers Possible and the Makers to Happen. I
doubt, however, whether Gregory's immaculacy of 'the
scientist' would be sustained by anyone today. Ivory
tower scientists claiming sanctuary for their discoveries
are rare today, and while there are many who shudder
away from the public debate many more recognize and
accept the social (and moral) responsibilities of science.
With the plenitude of new discoveries and the rapidi-
ty and scale of their application, scientific individuals

and societies feel as helpless as the rest of us in trying to curb the misapplication of these discoveries.

The change in the time-calibration, the drastic reduction of the lag between the conception of a scientific discovery and the technical application, is the main complication. Today there is no one to carry the red flag in front of the motor car. The feasibility study of a new development does not include the true evaluation of the social consequences or the possible erosion of human rights. We are hypnotized by gadgets or dazzled by spectacular achievements before we realize the encroachments they are making upon those rights.

Take the effects of miniaturization. One recalls that beyond the contraction which was necessary to put radar into aircraft an even more compulsive factor in miniaturization was the proximity fuse, the device which made a bomb or shell home in on a target, turning a near miss into a kill. The radar fuse which was developed had to stand 20,000 times the force of gravity when fired from a gun. It was a highly complicated radio set. To get results, the thermionic valve had to be reduced to the size of a vitamin capsule. It still had to be made of toughened glass, with filaments, grids, and so forth, and evacuated. The compact fuse in the nose of the shell had to include detectors, amplifiers, batteries, antennae, and firing switches, all on a midget scale. Similarly, the photoelectric proximity fuse, which made a missile 'see' its target, led to the development of tiny photocells of great durability and efficiency. It provided the research and development experience which was embodied in postwar television, the cinema, and industry. After the war we had the transistors replacing the thermionic valves by cold components now no bigger than a pinhead and requiring minimal power. With the refinement of printed circuits and solid circuits, the piece of furniture which had been the living-room receiver became the pocket transistor.

It is necessary to grasp the significance of this miniaturization because it is making possible reprehensible inroads on human rights and privacy. Size is important. Ancient Man could dig a moat against a mammoth but modern man is still vulnerable to a virus. Similarly, an individual can provide or demand protection against tangible or visible threats to his rights and take measures against invasions and intrusions *of which he is aware*—like demanding to see a warrant to search his home. It is difficult for any individual to take measures against the invisible or the inconceivable—the invasion of his privacy by electronic 'bugs' or by wire tapping.

Walls no longer ensure privacy. Distance no longer offers sanctuary. Darkness no longer provides a curtain and density is no longer a deterrent. Modern listening devices are manifold. They can be unobtrusively attached to a telephone. Transistor sets can fit into a

cocktail olive or be concealed in a lamp or in the up-
holstery of a car. They can be adapted to the filling of
a tooth or embodied in a denture. They can masque-
rade as wrist watches or tie pins. Stethoscopic microphones can be attached to a wall by suction or driven
into a party-wall as a spike. Directional microphones
like a sort of sound telescope can be used to pick up
conversations in an open space. The walking 'bug', a
person equipped with concealed microphones and
tape recorders, can sit at the next table, stand beside
you in a crowd, and eavesdrop a private conversation.

Modern viewing devices are also manifold. There
is the miniature camera no bigger than a cigarette
lighter; there is the 'casual' camera, no longer confined
to television jokesters; 'Peeping Tom' cameras which,
from a distance, zoom in on our private lives the way
that surveillance planes and observation satellites spy
on military installations of other countries. The infra-
red cameras can photograph in the dark. One-way win-
dows of Polaroid glass can enable viewers, unobserved,
to see clearly what is happening in another room.

Then there is the fibroscope. It consists of fine fibres
of glass, like the filaments in an electricity flex, which
enable optical light to be carried round corners. In its
beneficial form as the bronchoscope and gastroscope it
has revolutionized the medical examination of human
patients because this flexible light, which can travel
round corners, enables the diagnostician actually to

see what is happening inside the body. But the corollary of good is evil because this optical flex, just like electric wiring, can be strung into rooms to enable snoopers to see or photograph round corners.

It is dubious and debatable whether the police should ever be allowed to use such methods. Apart from that, with unrestricted availability of 'bugging' devices, the opportunities for the blackmailer, the common informer, and the character assassin increase. The community will become hagridden by suspicion, human relationships will be destroyed, privacy and peace of mind will be wrecked. Any mischief-maker could have his own do-it-yourself spy kit.

It is essential that the use of any such devices even by law enforcement officers and agencies should be rigorously restricted, with the public assurance that this kind of 'search' is never without warrant, given only in extreme circumstances. It is essential also that the sale, making, or possession of such devices for any purpose not fully disclosed shall be treated as 'criminal intent'; and that their use, without the knowledge of the party involved, shall be treated as a criminal act; and that the publication of material obtained by such methods shall be illegal.

Because we are bedazzled by new gadgets and seduced by the 'newspeak' about efficiency, without recognizing the portents, we are acquiescing in the usurpation of our rights. We are becoming the hostage of the computer. Anyone who thinks that the computer is merely a switched-on abacus or a glorified cash register had better call in a bio-engineer and have his own brain circuitry examined. It is no longer a docile machine which does sums. It has parahuman functions which surpass some of the faculties of the human brain which conceived it—for example, its superhuman memory capacity, its relentless logic, its speed of retrieval, and its tireless responses in supervision and control. It is constantly suggested, as some sort of reassurance, that the computer only gives back what a human being puts in. But it is not one human being; it involves many human programmers, who have disappeared in anonymity, in successive generations of computers. They have corrected each other's mistakes or, where the machine itself has rejected an illogicality, they have corrected that. In a literal sense, each generation of computer has inherited acquired characteristics. What is a matter of real concern in relation to human rights is its memory capacity—data storage. There is an increasing tendency of those in authority, for business reasons, or for purposes of government, to want to know more and more about each of us. For purposes of business, trading concerns contrive to find out as much as they can about our 'credit worthiness' and to share and to stockpile that information for their common benefit and protection. There have been government proposals for national data centres where all this public service information should be stored with telecommunications systems for retrieval. This sounds

engagingly plausible—just a national human inventory.
Everything would be there: health records, educational records, social security records, income tax records,
status records, and, of course, police records, including
the secret dossier one does not know one has. Dis-
parately, those records may be harmless, even beneficial. For example, a comprehensive health record, such
as would be possible in Britain with its National
Health Service, could be useful in an emergency; when
a casualty is taken into hospital in a strange town, it
would be possible to get his blood-group for transfu-
sion purposes in a matter of minutes from the computer centre. (As Lord Platt, former president of the
Royal College of Physicians, pointed out in the House
of Lords on 3 December 1969, it could be found out in
the normal way in five more minutes by pricking his
finger.) The computer might be an objective referee on
educational achievements and job qualifications but it
would not include genuine character references because no referee would be frank if it were going on the
national record. But even an official educational record
might be harmful to someone whose subsequent
achievements have obscured the fact that he never
made his school grades. The police dossier (unchal-
lenged because it is unchallengeable since the subject
does not know it exists) would now be reinforced by all
the information collected by efficient modern methods.
Item by item the national record may be inoffensive;

in aggregate it would present a misleading 'identikit'. In spite of all attempts to make the computer foolproof and knaveproof and promises that each area of information would be accessible only to those who have coded entitlement to have that information, there is no guarantee of the ultimate privacy of personal records thus assembled. As the Earl of Halsbury, President of the British Computer Society, said in the House of Lords Debate on 'Computers and Personal Records' (House of Lords, *Hansard*, 3 Dec. 1969), his own personal experience had taught him 'not to get starry-eyed about the efficacy of these security systems'. He quoted with her permission the views of Commander Grace Hopper, consultant-in-chief to the U.S. Navy: 'First is that those systems are proliferating throughout the United States everywhere and not one of them has not been beaten within six months by somebody clever enough to do so. Of course an amateur could not but the man who designs the system, or his assistant, can usually break it if he tries hard enough. Second, she has persuaded the United States Departments which rely on her for advice to grant to every member of every government department who has records on the computer an automatic right to a print-out. Anybody whose records are kept on a computer has the right to say "I want to see what the computer says about me". So if anything happens to change his status he knows that the computer has a record of it and can demand a print-out in intelligible form while there is time for a human memory to correct the thing if it is wrong.'

Lord Halsbury added his own strong views: 'You cannot send people to prison because the computer says so. The computer and the conventional system must run in parallel, at some extra cost . . . . If matters are to be kept secret, then they must be kept by the conventional system; if they are put on the computer then there must be the right of print-out. We cannot mix the two.' One of the basic human rights should be 'No secret information on a computer, and the right of print-out for the person to whom the computer records relate'. He insisted, 'These I believe are basic freedoms which are necessary for all who might easily become the victims of mistakes, let alone malice, unless we take precautions against it.'

Another measure of protection of the rights of the individual would be the legal obligation to inform him whenever it was proposed to use or divulge his personal data and for what purpose.

Inescapably it is necessary to admit computer evidence in the law courts because so many concerns keep their records on the computer and in one sense it is no more than calling for the ledgers. The party affected, however, must have the right to challenge the computer, satisfy himself about what was fed into it and that the machine was working properly. This, again, is no more than questioning the reliability of the human bookkeeper and the bookkeeping system. Courts must never get into the habit of accepting computer records as infallible.

It is essential to distinguish between statistical data and personal information. So much planning, social, commercial, and industrial, now depends on the availability of statistics that people find themselves being constantly interrogated and 'sampled' about their behaviour patterns, their attitudes, and their preferences. They have to be known to exist; that is, they are selected and identified by the data-collector but for the purposes of the survey they remain anonymous. If, however, a name or a number becomes associated with the statistic then it becomes part of a personal dossier and, in a different context, their admissions may be damaging or misconstrued. Conversely, we could all become depersonalized as numbers. The right not to be a number has been jealously insisted upon. A convict, as part of his punishment, became a number. One of the degradations of the concentration camps was the number tatooed on the skin. In the armed services, or in civil life in a state of national emergency, people became numbers but, in the past, concern for human rights had insisted on the restoration of their unique personality as soon as possible. As was pointed out in the House of Lords Debate (3 Dec. 1969), the functioning of government in such things as social security makes a national identification number an official convenience and, in a computer-based society, this could become the key figure, the common factor linking the various data systems. In the absence of rigorous safeguards, or in wrong hands or by mistake, it could be the means of total recall of the personal records of any individual.

The technological threats to human rights are not confined to 'bugs' and computers. There is such a thing as 'cultural privacy'—the rights of people within a nation to evolve and enjoy their own culture and not to have an alien culture imposed upon them. That raises the question of satellites and their uses. Existing communications satellites are ground-to-ground; that is to say, they reflect signals originating from a ground station to be picked up and relayed by existing ground systems. This process is selective, since local or national broadcasting services do not have to relay programs they do not want. Very soon, however, there will be direct broadcasting satellites, transmitters in the sky capable of beaming programs for reception in the homes. They could cover a whole region. The propaganda opportunities are obvious but so are the risks of cultural domination, the invasion of cultural privacy. A convention to ensure their proper use and to regulate their programming in advance of their siting would seem to be essential.

The protection of the personality in the light of advances in biology, medicine, and biochemistry raises questions more portentous than any concerned with devices or machines. We are hesitating on the threshold of the Biological Revolution, or one might call it the Bio-engineering Revolution because it implies that we can deliberately manipulate living components as we manipulate materials and machine parts for pre-conceived purposes.

I shall be dealing later with science and posterity but one must consider at this juncture what is involved for the contemporary individual, both for the medical scientist and for the patient who is, or should be, his concern.

A good example is organ transplanting. There would not, at first sight, seem to be much ethical difference between having a nose straightened or a face lifted by plastic surgery and having an internal organ replaced. Indeed, one could maintain that the internal organ was more humanly necessary than superficial cosmetics. If the transplant was by agreement between consenting parties (although often the 'consent' is post-humous) it might almost be regarded as a human right. But it creates an ethical dilemma of a different order. Even before kidney transplants people had an ethical unease about renal dialysis—the artificial kidney machine. Unquestionably it was a great technical advance, making it possible to treat kidney dysfunctions from which thousands die. The machine was, and is, expensive and involves intensive care of the patient by doctors and nurses. For whom the machine? In the United States the dilemma was evaded but not solved by having lay panels, like clinical juries, making life or death choices. In Britain, where the National Health Service entitles everyone, rich or poor, to have access to any necessary treatment, the responsibility rests on the medical staff. It was, still is, and will continue to be a difficult decision. Assume that some V.I.P., a famous man of advanced years, and a boy of, say, fifteen both require the only available machine, to whom would it be allocated—the living 'museum piece', whose obituary would thus be postponed, or a boy with an unpredictable future in which he might live to be a Nobel Prize winner or a criminal? A cynic would give you the answer.

With the discovery of chemical means of counteracting immunological rejections, it became possible to transplant actual organs. The natural defensive system, by which the body just as it can repel germs can reject alien, genetically incompatible tissue, could now be repressed so that the graft would 'take'. Thus a stranger's kidney could be transferred to a patient. Again, to whom a suitable kidney? That led to heart transplants and to heart and lung transplants. An essential condition of the operation was that the organs must be 'fresh'. To the point of scandal we have had a near Burke and Hare situation. (Burke and Hare were the Edinburgh grave robbers who supplied Knox, the University anatomist, with cadavers for dissection and finished up by providing warm corpses.) Doctors, collaborating with heart surgeons, would remove the heart at the moment of 'death' and rush it to the heart-operating theatre.

But what is the moment of 'death'? The old method of holding a mirror to the mouth to get the mist of breathing has long been discarded. The pulse, or the beating of the heart, is no longer the test. The test is the encephalograph to detect the brain waves. If the signal is weak in a hopelessly injured casualty then death is imminent and the patient expendable. If there is no signal, then the patient is dead. Thus the seat of life has become the brain. According to the signals of the encephalograph the instant removal of the heart can be determined and the organ rushed to the heart surgeon. Apart from the ethics of the decision, this situation raises the question: what is the ultimate object of the entire exercise? Assume (and it is a safe assumption) that any and every organ, presently, will be capable of being transplanted and that a human artifact could be reconstituted out of spare parts. To what end? Presumably, to service the brain, the vestigial remains of life. But the brain, however marvellous, is a biological computer. One can debate whether it is the 'mind'. It is certainly not the human 'personality'. Our personality is compounded from the chemistry and the responses of the whole body—not just the endocrines but our lymph glands and our gastric propensities as well. I have still not got a satisfactory answer from my organ-manipulating friends as to what would happen, for instance, to me, an entirely non-belligerent person, if I had a kidney from a pugalist complete with a suprarenal gland which provided the aggressiveness. We are not only remaking a machine of cells and tissues, we are remaking a personality. The question arises as to the identity of a person modified by surgical plumbing. What is the legal, moral, or psychiatric identity of a human so altered by medical manipulation that he has become an artifact?

There is another severely practical but profoundly important question: what is the cost in money, but also in human sacrifice, of the diversion of scarce facilities and scarcer medical manpower to the intensive care of those 'interesting cases'? Even those who get it are a tiny fraction of those who may need it. Thousands, tens of thousands, hundreds of thousands, and, in the wider world, millions of useful lives are being wasted, and an immense amount of suffering is being endured because of the lack of conventional treatment.

There is another dilemma. With all this 'doing over', with plutonium pace-makers being emplanted and attached to the heart, with the promise of miniature computers to be attached to our brain to compensate our failing memories, with electronic stimulators promised to enliven our 'pleasure centres', with all the advances in biochemistry and geriatric care, and with the protection against and treatment of infectious diseases, we can extend the span of life. We can keep people alive indefinitely. This is a social problem of the first magnitude. It is also a personal problem for all of us because it involves a basic human right—the right to live or die with dignity.

There is a need for a consensus of conscience; an international tribunal of ethics; a surveillance (which need not be suppression) of the kind of things which are happening; and a new personal ethic for doctors and medical scientists. The Hippocratic oath, noble in its intention, is no longer adequate. It has been overtaken by events. The doctor, under the oath, pledges, 'The regimen I adopt shall be for the benefit of my patients, *according to my ability and judgement* and not for their hurt or any wrong.' Today, his abilities far exceed his judgement because medical science and technology have given him means to treat or operate on conditions which would have been intractable a few years ago. Indeed, he could construe his Hippocratic oath as making it incumbent upon him to apply any knowledge he has down to the most recent discovery reported in the latest medical journal or the newest drug offered him by a pharmaceutical firm. In present circumstances it is possible that he might be sued for professional incompetence or malpractice for withholding an up-to-date form of treatment. In the United States a medical practitioner was successfully sued for damages for not prescribing a drug of which he had never heard. The 'judgement' which innovation imposes on a doctor exceeds the professional common-sense on which the 'good doctor' would rely in good conscience. In the phrase of the biophysicist Leroy Augenstein, 'He is being asked to "play God".'

In the compassion of true medicine, human dignity cannot accept the bio-artifact, the non-person, the living-dead. I, personally, insist on the human right to say, in full possession of my faculties, the conditions under which I do not want to be kept artificially alive.

# III

## SCIENCE AND POSTERITY

How LONG have we got? Is *Homo sapiens'* lease of this Planet Earth really running out?

In his bleaker moments, Bertrand Russell said that it was a fifty-fifty chance that the human race would not survive this century. That was when he was convinced that a nuclear war was inevitable and that devastation and radiation would destroy life and leave this planet a radioactive desert. That was what Brock Chisholm meant by man's capacity to 'veto the evolution of his species'. If we avoid a total nuclear war, there is the other grim alternative that the species may multiply so much and so fast that it will starve or stifle itself to extinction. This includes the prospect that Space Capsule Earth will become a Black Hole of Calcutta in which too numerous and too congested human beings will die from the heat generated by their own bodies.

We have to remember that Planet Earth is indeed a space capsule which is travelling at nineteen miles a second in its orbit round the sun. It is enclosed in an envelope, the atmosphere, strapped to it by the forces of gravity. While astronauts have demonstrated that it is possible to breach the gravitational fence and escape to the moon, for all but the very few, we earthlings are confined to the surface of this planet and our living space is definitely limited.

An eye, human or electronic, looking eastwards from the moon or from a space vehicle sees a globe round which a man-made satellite can travel sixteen times a day and of which only three-tenths is land and seven-tenths is covered with water. That eye, human or electronic, sees mountains reduced to wrinkles; evidence of surviving forests; tawny expanses of hot deserts which cover a fifth of the earth's surface and cold deserts which cover another fifth. It can distinguish the pattern of cultivation, the arable or cropped lands which account for only one-tenth of the land surface. That is the extent of man's family estate of which we are, at present, the improvident stewards.

In those terms, earth is not, like the moon, the inert geological relic of a cosmic incident; it provides the biosphere, the living space, for the evolutionary process which, we like to think, had its consummation in man.

A quarter of a million years ago, or thereabouts, *Homo sapiens* took a lease on Planet Earth. He shared occupancy with his fellow lodgers: the beasts, the birds, the fishes, the micro-organisms, and all the forms of plant life. His, however, was a special contract with nature, because he was now Thinking Man, whose conditions for survival depended on the use of his intelligence, his imagination, his foresight, and his finger-skills.

Let us be clear about this when we speak about man 'interfering with his environment'. The only reason we are still here is that *Homo sapiens* managed the hostile and inhospitable environment. The odds were decidedly against him because, without fang or claw, beak or talon, fur or feather, scale or carapace, naked to his enemies and to the elements, he had been, of all the creatures of the earth, the least likely to survive. He could not outwrestle or outrun his natural predators but he could outreach them with clubs and spears and slings. He made fleshing-tools to strip them of their pelts and clothe his nakedness. He tamed fire, which terrified all other creatures, and used it to heat his caves and cook his food. To that extent, he mastered his environment because with clothes and heat he could escape climatic restraints. In this way he had leased the whole planet since he could choose to migrate and settle anywhere from the Tropics to the Arctic.

Later, from being a hunter, he became the domesticator of animals—decoying, taming, and breeding them. With such flocks and herds, pastoral man moved to seasonal pastures synchronizing with nature. Then from being a food gatherer, he became a food grower. He discovered that seeds and certain grasses were nourishing but he also discovered that, if those seeds were scattered, they would take root and grow and that they would grow better if the soil were delved to receive them and better still if they were properly watered. So he became a tiller and irrigator, settling in the alluvial plains of the rivers and creating a self-sufficiency of food for his family and for his domesticated animals. He presently found that there was efficiency in the division of labour and that some of his kind were more proficient than he or others in contriving tools, making pots, building houses, weaving baskets from rushes or clothes from finer fibres, or, a great advance, making wheels to lighten his haulage. From the produce of his land, the tiller obtained the produce of the craftsmen.

As a husbandman, he was still in communion with nature, nursing the soil, but he provided for the priesthood which had to intermediate between him and the gods who were the embodiment of elemental nature, sending thunder and lightning, floods and droughts, pestilence and earth convulsions as reminders to Thinking Man that, no matter how ingenious he might be in modifying his environment, there were two parties to the lease and nature was still the landlord.

The temple tithe-barns of the farmers' tributes to the gods became entrepôts. The traffic in those farm goods produced trade with other communities and, when direct barter became too cumbersome, tokens of exchange had to be created; so money was invented.

The money became a commodity in itself to be traded by the money-lenders, the money-changers, and the merchant bankers. Tangible wealth became attractive to marauders. So protection was necessary and warrior kings and conquering dynasties came into being. Settlements became walled cities.

This, the growth of cities, we call civilization but the whole superstructure, the craftsmen and their auxiliaries such as miners, the priesthood, the tax gatherers, the bazaar traders, the import-export merchants, the financiers, the soldiers, and the feudal systems and imperial adventuring of kings all depended on the labour of the tillers and on the soil which they husbanded.

Cities brought their own hazards. Thousands of years ago there were problems of overpopulation; congestion within the city walls; slum conditions; infections and contagions; sewage and trash accumulations, so that the archaeologists find the miniaturized version of our present pollution problems in the succession of cities built on the midden heaps of their predecessors.

And there were conservation problems. For irrigation, dams had to be built. The greatest king of the First Dynasty of Babylon was Hammurabi (circa 1800 B.C.), who promulgated a legal code regulating the management of dams and canals. For instance, Section 55 reads: 'If anyone be too lazy to keep his dam in proper order and if the dam breaks and all the fields be flooded, then he in whose dam the break occurred shall be sold for money and the money shall replace the crops he has caused to be ruined.'

In the excavation of the Indus civilization of 5,000 years ago, the ruins of Mohenjo-Daro revealed the existence of booths where drinking water was sold. Alongside were huge spoil-heaps of broken cups, far in excess of the ravages of the most careless dishwashers and evidence of a sanitary code: When someone bought a cup of water, the crock was broken. This is what we would applaud as 'hygiene'. Today, and every day, hundreds of millions of plastic containers, cans, and non-returnable bottles are discarded. In the cities we call this 'refuse'; in the countryside we call it 'litter'; and, in general, we call it 'solid pollution'. The once-onlies of the Indus civilization have thus become a threat to ours. The largest kitchen middens of neolithic man, accumulated over centuries, measure no more than a few thousand cubic metres. Any township deserving a mayor will produce that much in a week.

Four thousand years ago the Minoan civilization was advanced in sanitation. Clean water was brought to the capital, Knossos, in pressure pipes; cemented stone drains provided the sewerage system; dry refuse was collected and deposited in large pits outside the city, earth silos in which wastes fermented into compost for manuring the fields. In A.D. 100 nine great aqueducts, with a total length of 424 kilometres, brought water to Rome and lead pipes carried it to houses and to one thousand public baths. The *Aedile* (the nearest approach to our borough engineer) could ride in his state barge along the main drain, the *Cloaca Maxima*, under the city to the Tiber. His scavengers, slaves or prison convicts, shovelled the city's refuse, domestic and animal, through manholes into the sewer to be flushed into the Tiber and thence to the sea. This was commendable public health engineering. Today, grave concern is expressed with justification about the pollution of the Mediterranean by untreated domestic and industrial sewage. The sea itself has become the *Cloaca Maxima*.

In 1229 Henry III granted the first charter to the town of Newcastle-upon-Tyne to dig for coal for the comfort and warmth of its citizens. Seven hundred years later citizens of London, by the thousand, were dying of lethal effects of coal-burning. Today, by imposition of smokeless zones, London's public buildings are being restored to their pristine state.

In 1859 Drake's Well was drilled at Titusville, Pennsylvania, and the internal combustion engine was on its way to unleash ill-combusted hydrocarbons from automobile exhausts into the atmosphere. Today 6,000 million tons of carbon dioxide ransacked from the geological vaults of coal, oil, and natural gas, which were locked up tens of millions of years ago, are being vented into the atmosphere as an *annual* increment to the 360,000,000,000 tons of carbon already released in the industrialization processes of the past century. The concentration of carbon dioxide in the air we breathe has increased by approximately 13 per cent above the equilibrium of a century ago, and if all the known reserves of coal, oil, and natural gas were to be used similarly, the concentration would be ten times greater. In other words, the amount of carbon dioxide would have been doubled.

This increase is something more than a public health concern about damaged lungs and smarting eyes; it can materially affect the climate and conditions of the biosphere, on which all things, including humans, depend for survival.

The climatic changes involve the 'greenhouse effect'. This is not a new scientific discovery. It was first suggested by William Tyndall in 1863.

The atmosphere is semi-transparent for solar radiation but is fairly opaque for terrestrial radiation—that is, for the heat radiated by the earth itself. For instance, in clear atmosphere (containing an average amount of water vapour, carbon dioxide, ozone, and dust) about 65 per cent of the incoming solar radiation reaches and is absorbed by the earth's surface. For the same atmospheric conditions, however, only about 10 per cent of the total radiation leaving the earth's surface is directly released into space. The rest is reflected back to the surface. As you can see this is similar to a greenhouse which lets in the sun's rays but confines the heat under glass.

In this mechanism, carbon dioxide is important because of its strong absorption (and subsequent emission) of infra-red radiation. In the atmosphere it would absorb the earth's heat, re-radiate it back to the surface of the earth, and further inhibit radiative cooling of the ground.

Fifteen years ago, when I was in the Canadian North and before the International Geophysical Year surveys began to quantify the carbon dioxide effects, the process was recognizable. It was beginning to impinge on the permafrost system. The Old Hands in the Subarctic were commenting on the extension of the black earth northwards. Areas which had been permanently frozen were beginning to thaw. In the Barrens the snows had reflected the sunlight but the seasons seemed to be changing. The black earth of the thaw was left longer exposed and black earth absorbs and retains the heat of the rays which the white snows would reflect. Thus the warmer soil remained longer exposed. In Norway they began to find in the black earth of the erstwhile snowlands the arrowheads of the time of Eric the Red, a thousand years ago, when in the natural cycle the soil had been last exposed.

You have probably heard the people who say that the scientists cannot make up their minds; they say that carbon dioxide will heat up the surface of the earth while at the same time it will cut off the sun's rays. 'Make up your minds,' they say. 'Are we going to die of heat or die of cold?' Of course, it is not like that. We are talking about a system in which the greenhouse effect can operate at ground level, as it were, while upstairs in the stratosphere, at 25 kilometres and beyond, a different process is working. The ozone (oxygen$_3$)

absorbs solar radiation, raising the temperature at such altitudes, but carbon dioxide cools it by borrowing part of it so that a drop of temperature results in the stratosphere. The nature of those relationships, the heat exchange between ground heating and a stratospheric cooling, changes the transport system and the role of evaporation and condensation and the role of the ocean in providing a heat reservoir. The climatic forces are, therefore, altered and the lows and highs of the weatherman's chart become childish doodles by comparison. A doubling of the carbon dioxide concentration over the present level, which is possible if we accept the estimates of demands and use of fossil fuel by a doubled population, would result in an increase in mean surface temperature of about 2 degrees centigrade and possibly a decrease in the stratosphere temperature of 4 degrees centigrade. This result will radically change the weather system. Areas now fertile will become arid; arid areas will become pluvial. But where? If the man-made glasshouse effect at the surface of the earth operates on this doubling principle, ice will begin to melt. The polar ice packs are not going to raise the sea level because, as good old Archimedes assured us, the weight of water displaced is equal to the volume of the solid. So floating ice does not count but land-locked ice does. The Greenland and Antarctic icecaps in thaw would be supplemental in volume and would raise the levels of the oceans, altering the fretwork of our coastal land masses. (I have been advising my friends not to take ninety-nine-year leases on sea-level properties.) Moreover, there are the ice dams. Those in the Himalayas and elsewhere are holding back stored waters and, if melted, would wreak havoc in the lower levels. However sceptical one may be about the estimates and calculations, the excesses of carbon dioxide, man-produced, cannot be ignored. In greater or less degree it has much to do with man's real-estate options. The International Geophysical Year of 1957/58 alerted us. Far more scientific research is needed to inform us.

And now we have become aware of the ocean. (We talk about the seven seas but actually the ocean is one.) Seven-tenths of the earth's surface is drowned beneath the waters, an area so vast that it seemed that puny man could not possibly threaten it. We have sailed over it. We have hunted fish and sea creatures, including whales, to the risk of extinction. We have dumped, as far away from our own backyards as possible, deadly materials: high explosives, poison gases, radioactive materials, dangerous industrial chemicals, trash of all kinds, and domestic and industrial effluents.

We might, if we thought about it to that extent, console ourselves that nature has been pretty extravagant in the way it has discarded stuff into the oceans. For countless millions of years, the oceans have been receiving substances which we nowadays categorize as pollutants: radioactive nucleotides as a result of cosmic ray bombardment from outer space, and the erosion of rocks containing radioactive elements; the erupted products of volcanoes and chemically charged effluvia from seismic vents; hydrocarbons such as those found in oil, coming from natural submarine seepages, due to earthquake ruptures or evolving from the natural decay of marine plant and animal life; the accumulated effects of hydraulic mining—the natural erosion by rivers and coastal waves and (always underestimated) the wind-transported particles from the continental hinterlands.

For example, for thousands of years, around the Mediterranean, man's 'civilized' activities spilled pollutants. In the samplings of the sea bottom of the Mediterranean one finds the discards of the earliest copper-miners in Cyprus, of the ferrous metal-workers of Asia Minor, of the Phoenician tin-workers, of the Hibernian mercury-workers, and so on. And, of course, there was always domestic sewage but that encouraged, rather than threatened, marine growth.

The difference between now and the past, remote and more recent, is that human activity, multiplied by numbers, the needs and the ingenuity of an enormously increased population, is adding substantially to the amounts of those materials and particularly of man-contrived substances such as chlorinated hydrocarbons and radioactive materials which did not exist in nature and with which the marine ecology cannot

cope. Such activities, divorced from nature, are doubling the natural concentrations of marine chemicals
and introducing new chemicals—innovations—in quantities approaching those of naturally occurring chemicals. Those concentrations may be localized in terms
of the vast expanses of the oceans but the effects may
not be localized.

The direct effect of surplus and novel chemicals on
the marine environment, particularly within enclosed
systems such as the Mediterranean, is on marine biology. Marine life is a fine web of interrelated food
chains which depend upon the chemical constituents
of the sea. Diversity of species is essential to the sta-
bility of the ecological system. Crucial to the food
chain is the phytoplankton which are responsible for
90 per cent of the living material of the sea and, moreover, for about 70 per cent of the oxygen of the earth.
That is, our breath of life depends upon them. They
provide the 'pastures' for the rising scale of sea creatures. Their function and their profusion reside in the
sea's surface layers to depths penetrated by the sun's
rays. Those rays are responsible for photosynthesis, a
process which, apart from producing food, liberates
oxygen.

The ecological balance of the complex chain of life
can be upset in a variety of ways. Some pollutants
simply poison the plants and animals with which they
come in contact. Others make such demands on the
oxygen dissolved in the sea water that other living

things competing for that oxygen suffocate for lack of it. Some pollutants encourage the excessive growth of a single species of plant or animal so that it prevails over others. Other pollutants are concentrated in species which have an affinity for them, without deleterious effects on themselves, but which pass them on up the chain in increasing doses until they become dangerous or lethal to other species, including humans.

An example of the first—direct poisoning—was the consequences of the spilling of 200 pounds of endosulphan into the Rhine, with the massacre of the fish life. An example of the second—oxygen deprivation—is oil pollution. Apart from killing sea birds and so on, the natural decomposition of an oil slick requires its oxidization by the action of bacteria in a process which depletes the dissolved oxygen supply on which marine life depends. (One litre of oil depletes 400,000 litres of water of its oxygen.) An example of the third—excessive growth—is eutrophication arising from both domestic and industrial discharge which provides an excess of nutrients such as nitrates in fertilizers or phosphates in detergents. The excess produces a 'bloom', a rapid growth of responsive marine species, usually phytoplankton, which multiply at the expense of other species. This is familiar in the coarse, slimy, blue algae associated with the outfalls of untreated sewage or the 'red tides' of the Florida and California coasts and elsewhere. Obnoxious smells are usually associated with such 'blooms', which in their proliferation kill other forms of life, and the blooms themselves, when they die, cause the deoxygenation of the water and the production of a sea desert. An example of the fourth—the 'addiction' of certain forms of life for certain chemicals—is the now familiar story of D.D.T. This chemical can directly kill insects by paralysis. One of its virtues in the prevention of insect-borne diseases is its persistence, lasting decades. It has been estimated that about 1,000 million pounds of D.D.T. have already been introduced into the biosphere, and what with the wind-borne aerosols of spraying and the run-off from agricultural activities and D.D.T.'s long-life, most of that amount has, or will, finish up in the oceans. D.D.T., like other chlorinated hydrocarbons, is not readily metabolized but is stored in fat. Even when it is metabolized (changed by physiological chemistry in the body itself) the end result is another form of chlorinated hydrocarbon. Thus D.D.T. (and the other chlorinated hydrocarbons) accumulates in marine life and is concentrated, in increasing proportions, in the food chain. Oysters have been found to amplify small concentrations of D.D.T. 70,000 times in one month. D.D.T. has been found in penguins in the Antarctic, where D.D.T. has never been used and thousands of miles from any sprayed areas. It has been found in the flesh of whales off Greenland.

Although we have seen what has happened to the

streams, the rivers, and the Great Lakes, we still tend to think of the oceans as so vast that they can cope with all the waste we like to put in them. We forget that the ocean is a living system, not just in terms of the plants and creatures which it sustains within it but in terms of its circulation. This fact was powerfully brought home to me in the study which I did on the pollution of the Mediterranean.

The circulation of water, and with it of oxygen, in the Mediterranean is regulated by two 'lungs' located in the Provençal Basin and the Upper Adriatic. The currents of Atlantic water, coming into the Mediterranean through the straits of Gibraltar, sweep into those two localities where the 'healthy', oxygenated, surface waters are chilled by the cold winds spilling down from the Alps. They become dense and sink. The water thus transferred to the depths off the south coast of France goes into circulation in the western Mediterranean; that from the Adriatic goes over the sill of the straits of Otranto and travels eastwards to the Levant. With urbanization, industrialization, oil shipments, and the run-off from 'medicated' soil of the inland farms, it is just in those lungs that the major pollution of the Mediterranean is taking place. If the transfer of oxygen between the atmosphere and the surface of the sea is impaired by thermal pollution or chemical layers or if the oxygen released by photosynthesis in the upper layers is reduced by interference with the micro-organic life, the respiration of the

Mediterranean is affected and ultimately it will become a stagnant sea, like the Black Sea, depleted of oxygen. This possibility, in terms of posterity, is something more serious than the inconvenience to present-day tourists of filthy beaches. By our irresponsibility today we are reducing the options. We are mortgaging man's family estate and nature may foreclose.

I have said, in the House of Lords, 'Pollution is a crime compounded of avarice and ignorance.' Avarice because of the reckless use of resources and destruction of amenities and the environment for the sake of quick profits. Ignorance because industrialists do not bother to find out and anticipate the effects of their activities. But we who protest against such things are also accessories to the crime. If more and more people want more and more goods and want them cheap, then the abuses become more and more exaggerated. It is a question of cost. Pollution can be prevented by industrial precautions, which are costly. When we insist upon them, industrialists tell us, 'But you will price us out of business because our foreign competitors are not under those restrictions.' Or the community picks up the social costs. But I also try to remind businessmen that pollution is gross mismanagement, incompetence, and the squandering of real wealth. What we call fumes or effluent or slagheaps contain valuable materials. Even eutrophication, which we hear so much about, is too much nutrient (from domestic or industrial sewage) in the wrong places.

Suppose that by improving our behaviour as tenants we get an extension of our lease of Planet Earth, what sort of 'posterity' do we see frequenting our secular Hereafter? There are some people who would be prepared to write out the prescription now. Hitler's Reich, which was to last a thousand years, was to be peopled by blond Aryans by the simple expediency of eliminating or discouraging the propagation of unwanted strains. With better intentions, and with more recent insight into genetic methods, there are those who would seek Eugenic Man by manipulating hereditary traits to remove those which are deleterious.

To me the DNA, or Bio-engineering, Age is even more portentous than the Atomic Age. We know what we did with the Secret of Matter, the energy within the nucleus; we exploded it as an apocalyptic bomb. Now we are dickering with the Secret of Life with no real understanding of the moral and ethical, or indeed the practical, consequences.

Let us consider tomorrow's children. There is not much doubt that we shall be able to prescribe the offspring in advance. The fertility drug, fortuitously, has shown that it is comparatively easy to multiply the number of children from one conception—human litters. There are the established techniques of artificial insemination by which an infertile husband agrees, prenatally, to be the foster father of the child his wife will bear from an unknown man. With sperm banks male genes can be kept indefinitely. It is only a matter of time before unfertilized or fertilized ova will be similarly preserved indefinitely. It is already possible—indeed a growing practice—to take the fertilized ovum from one female animal and transplant it into the womb of another female animal, a prenatal foster mother who has contributed nothing to the genes but who will gestate the offspring and deliver it. With such devices it is possible to juggle with time. Imagine the genes of William Shakespeare being mated with those of Mrs. Siddons and parturated by a Hollywood starlet! Imagine the complications in the hereditary peerage of the offspring of the 10th Earl being born after the 13th Earl! It is now the practice in the United States to preserve the sperm of a man who decides to be sterilized so that if he loses his existing children his wife could replace them. There are serious suggestions that fertilized ova, the parental genes of which are all spelled out, could be available like seed packets (possibly with pictures of the expected offspring on the packet) so that women can choose the babies for which they would be artificially impregnated. Imagine the prescription: red curly hair; blue eyes; six feet tall; athletic; musical attributes and high I.Q.; and the genes would be dispensed accordingly.

Joshua Lederberg, Nobel Laureate, foresees with alarm and not approbation the possibilities of 'cloning' within the next fifteen years. Cloning means that it will be possible to grow from the nucleus of an adult cell a new organism which will have the same char-

acteristics of the person contributing the cell nucleus.
The resultant human copy would start life with the
identical genetic endowment as the donor, indis-
tinguishable in every characteristic. They would be
more identical than 'identical' twins. Their physical
appearance would be like dolls mass-produced from
the same mould. Their mental processes would be the
same, so that they would have the same psychic awareness, which would amount to telepathic identity.
They would synchronize their responses. What a wond-
erful ball team they would make! But how could you
recognize the goal scorer? Cloning has already been
artificially produced in plants and in amoeba, and
Lederberg is convinced that 'somebody is doing it
right now' with human cells. This process would play
ducks without drakes with the evolutionary processes.

Another upheaval probably more imminent is sex
determination, which would enable parents to determine in advance the sex of the child. Such a prospect
does not sound unduly alarming except to demo-
graphers. What would happen to the pattern of population if a biological Christian Dior persuaded women
that boys are all the fashion? But there is a 'prescription for posterity' so fundamental that it transcends all
such considerations. That is DNA.

Away back in 1953, as science editor, I induced my
newspaper to use a massive one-word headline. It was
*Deoxyribonucleicacid* and I started off the article: 'Deoxyribonucleicacid. Deoxyribonucleicacid. Deoxyribonucleicacid. Repeat it. Spell it, and repeat it again. It

is one of the most important words you will ever have to learn.' I would not say that it is a household word, or a barroom word, even today but, as DNA, it is well enough known to worry a lot of people.

In 1953 Watson and Crick at Cambridge and Wilkins at King's College, University of London, unravelled the strands of the double helix and qualified for the Nobel Prize. They were studying the molecular structure of DNA, the basic chemical of the living cell. They determined that it was a double helix, like a spiral staircase. They found that the nucleotides were arranged in antiparallel chains. (Antiparallel like the 'up' and 'down' lines of a railway track.) The chains were cross-linked by hydrogen bonding. It was, however, a peculiar kind of spiral staircase. The spiral strands were linked groups of sugar, phosphate, adenine, thymine, cytosine, and guanine. The last four were critical components, which combined with the others formed nucleotides. The nucleotides, identified by their initials A.T.C.G., lined up as pairs linked by hydrogen bonds, which formed the rungs of the staircase—A.T., T.A., C.G., and G.C. It made an essential difference which nucleotide was on the right and which was on the left. Whatever the arrangement, it would in cell replication go on repeating itself like the patterns on wallpaper.

Here then was an insight into how information is conveyed from cell to cell, or generation to generation, dictating what will grow out of it—a spinach cell does not produce a George Bernard Shaw. Each minute cell must have millions of bits of information specifying 'Be a brain', 'Be a curled hair and not a straight one', 'Be a black eye and not a blue one'. Here, in the arrangement of the nucleotides, was the chemical code.

Remembering the twenty-six letters of our alphabet which are necessary to form words in our dictionary, it is difficult to imagine how to get along with four letters, A.T.C. and G., but think of the amount of information which can be conveyed by the dot, the dash, and the pause of Morse code. It is estimated that in the forty-six chromosomes (the packages of genes) of a fertilized ovum there is contained five thousand million nucleotide pairs. The genetic specifications thus encoded for producing a person from the fertilized ovum would, if spelled out, in English words, fill five hundred volumes of the size of the *Encyclopaedia Britannica*.

The important thing to remember is that the code can be altered by varying the position and disposition of the nucleotides and the new version will go on repeating itself generation after generation. This process is what is called 'mutation'. We heard a lot about mutations after the atom bomb went off, but it was known as long ago as 1927 that mutations could be produced by high-energy X-rays. This can be regarded as breaking the DNA code by physical force—jerking the nucleotides out of position and causing them to regroup—like splicing a broken film in a wrong sequence of frames. But many chemical substances are capable of changing the genes which determine our hereditary traits. Nitrous acid, for instance, can affect a nucleotide in such a way that its hydrogen bonding is changed and different conjunctions of the DNA 'alphabet' occur. The effect of this manipulation of the genes by giving them different DNA instructions can drastically alter the nature of cells and of the next generation.

It is true that with increasing knowledge of specific elements in the code—like deciphering a secret code in espionage—we would produce beneficial results. We could change the messages to cancer cells and make them harmless. We could recognize and avert congenital defects even in the womb. We could change the genetic instructions which at present produce hereditary diseases. We could alter the coding of plants so that they would yield more food. But we could produce man-made viruses and bacteria for use in biological warfare against which there would be no immunological protection. The nature of men and women could be changed.

Who will write such prescriptions? Not the original discoverers nor the Nobel-winning successors. They are dismayed by the possibilities of what they produced in pursuit of human knowledge and have turned to other researches, saying in one way or another 'Do not tamper with heredity. The mistakes will be irreversible.' But knowledge once given cannot be taken back. There will be others who will try to write prescriptions for posterity, spelled out in the DNA code. There must be another kind of code—a code of conduct for those who tamper with codes.

The medieval alchemists (actually much maligned) were accused of Black Magic. Their propensity for evil was far less than the best-intentioned scientists of today because they are too innocent to foresee the abuse of their discoveries. Einstein, whose equation $E = mc^2$ spelled out the release of atomic energy, said, when confronted with the horrific consequences of the Bomb, 'I wish I had become a blacksmith.' Scientists are the trustees of knowledge but in the words of the Latin tag *Quis custodiet ipsos custodes*? Who takes care of the caretakers? The answer is 'No one'.

Science, I repeat, is a human right. Unless one is superstitious there is nothing wrong with natural curiosity. It is what is done with the knowledge inquisitively acquired which creates the difficulty. For knowledge is not wisdom; wisdom is knowledge tempered by judgement. We go back to David Hume's 'is' and 'ought'.

Sir Macfarlane Burnet, who won the Nobel Prize in 1960 for his work on tissue transplants and who developed a crisis of conscience about the possible applications, said: 'It seems almost indecent even to hint that, as far as medicine is concerned, molecular biology may be an evil thing.' The scientists who won Nobel Prizes for DNA have similarly recoiled. James Watson appeared before a committee of the U.S. Congress to

urge that restrictions be imposed on human cell manipulation. Crick, confronted with the wilder genetic
possibilities, urged people 'not to stand for them'.
Wilkins became president of the British Society for the
Social Responsibility of Science. George Beadle
pleaded with his colleagues in the DNA field not to
lend themselves to gene manipulation because 'the
effects will be irreversible'. Salvador Luria declared
that 'geneticists are not yet ready to conquer the earth,
either for good or evil'. Max Perutz warned against
fertilizing human eggs in test tubes because of the
great and quite unjustified risks of producing a de-
fective embryo. One Nobel Laureate after another has
served notice. The operative phrase is 'unjustified
risks'.

Somehow we must bring together a body of wise
men from all over the world and from diverse philoso-
phies and cultures to consider the inventory of opportunities but also of mischief, actual or potential.
They must give us the basis for establishing the 'is' and
the 'ought'. They must produce instruments and institutions, legal and professional, which can apply ethical
restraints. They must know, but in that knowledge
they must judge in the best interests of mankind.

How long have we got? Well, it is later than most
people think. The alarm clock has gone off but, as
parents of young children know, the child climbing
up on your chest or poking a finger in your ear can

rouse you more effectively than the bell. The youth of today and the protesting students are doing the Dennis the Menace job of clambering over our somnolent forms and sticking their fingers in our ears. We can make it if we rouse and run, but we have to run because we are on a down escalator and must move fast just to stand still, let alone mount. Our difficulty is how to arouse many of our institutions, including our academic institutions; how to arouse our political institutions and our institutions of moral evaluation, including the churches. The churches have the Sodom and Gommorah syndrome: if disaster is coming it is because we deserve it. Instead of standing back and disapproving, with pious horror, the self-assertiveness and irreverence of the young, we should be listening to their concern and their recognition of the unpleasant truths of the human predicament, which we, not they, created.

In the 1950s, young people marched to the dirges of despair because, like Bertrand Russell, they believed there was no future; the Bomb was to take care of that. Then, in the 1960s, when they began to realize that they might have a future, they did not like the look of the future they foresaw. Now young people are beginning to identify the nature of that future. Their concern about pollution, their concern about the environment, and their recognition that non-renewable resources are being squandered have brought them to their point of modern heresy. They are daring not only to question but to deny that growth, either for profit or for gross national product, is a paramount necessity. They recognize that prosperity bought at the price of pollution, of built-in obsolescence, of non-returnables, is no betterment of the human condition and is ominous for their future. Growth to meet growth in terms of human essentials, yes. Growth for the sake of growth itself, no.

In my experience the younger generation is asking the right questions. They are setting a timetable by the very urgency of their protest. That protest may sometimes sound discordant. It may be in an idiom that their elders find distasteful. Their methods may be brash and often repellent. But they are sticking their fingers in the ears of those who did not rouse to the alarm clock.

Their priorities seem to me to be right and even unselfish—because it is very easy to conform and become Organization Man, especially when professional jobs are scarce and becoming scarcer.

After all, it is their world we are talking about. It is their responsibility from now on and the least, indeed, the most, my generation can say to theirs is 'We shall do everything possible to ensure that we do not do anything more to reduce your options to make your own mistakes.'

Time is of the essence. If we convert our awareness and our concern into institutionalized efforts to change the trends, we can postpone perhaps indefinitely the forbidding prospects on which these lectures have dwelt. I am an optimist. I believe we can beat the clock because I sit at the feet of my own children and listen with profit and encouragement.

One last word. I have given up predicting; I now prognose. There is a difference; if you predict, people plan for the prediction and in so doing confirm the trend. In prognosing, one is saying, like the wise physician, 'I do not like the symptoms, and if you and I do not do something about it your prospects are pretty grim.' But the operative phrase is 'if you and I do not do something about it'. Here is the option and the commitment and the challenge which removes the sense of inevitability and increases the time span.